Fábio de Souza Ferreira

Geometric mathematics in the slab cistern

Fábio de Souza Ferreira

Geometric mathematics in the slab cistern

Learning mathematics through the construction of a slab cistern

ScienciaScripts

Imprint

Cover image: www.ingimage.com

This book is a translation from the original published under ISBN 978-613-9-67453-4.

Publisher:
Sciencia Scripts
is a trademark of
Dodo Books Indian Ocean Ltd. and OmniScriptum S.R.L publishing group

120 High Road, East Finchley, London, N2 9ED, United Kingdom
Str. Armeneasca 28/1, office 1, Chisinau MD-2012, Republic of Moldova, Europe
Printed at: see last page
ISBN: 978-620-8-16806-3

SUMMARY

I dedicate this work to my wife and son, who gave me the support and strength not to give up. I dedicate it to my mother, who gave me the opportunity to live.

ACKNOWLEDGMENTS

I would firstly like to thank my wife and son for their patience and understanding during the countless times I had to spend late nights writing reports. I would also like to thank Prof. Me. Thiago Jefferson de Araùjo for his patience and collaboration in guiding me.

To all those who believed in this work and gave me the strength and encouragement to continue with this project.

To God and his son Jesus, creator of the heavens and the earth, who allow us to live our lives with greater peace, faith and, above all, divine protection.

"There are no easy ways to solve difficult problems."

(René Descartes)

SUMMARY

This work is the result of a Research Project for a Course Conclusion Paper (TCC) developed in a rural community called Sitio Barra do Japi, which is located in the transition between the cities of Japi-RN and Santa Cruz-RN, both situated in the Agreste Potiguar mesoregion and Borborema Potiguar microregion, about 122 km from the capital of the State of Rio Grande do Norte, in Brazil. The aim is to investigate and analyze through ethnomathematics how farmers and their closest relatives can learn mathematics during the construction of a plate cistern. To do this, we tried to bring together the mathematical concepts involved in the construction work. The aim is for the community's math teacher to be able to integrate his lessons with the daily reality of these people, trying to bring the school closer to the daily lives of the community's residents, getting the students to take part in activities that relate to the mathematical teachings developed and based on the concepts of Ethnomathematics. The aim is also to bring them closer to the mathematical knowledge that rural people bring with them without ever having attended school. To this end, historical and socio-cultural facts related to these people's way of life will be investigated. The emphasis will be on learning mathematics through activities directly linked to the cistern. However, the methodological appropriation of the concepts and techniques of Ethnomathematics based on the thinker D' Ubiratan Ambròsio will be necessary. Finally, we intend to show the results of these analyses through statistical studies from the intervention carried out in the community.

Keywords: Plate cisterns.Education.Ethnomathematics. Teacher.

1 INTRODUCTION

Faced with the new challenges of teaching mathematics, our intention is to mitigate the negative effects of traditional methods. We will also seek to bring theory and practice closer together, based on the fact that mathematics is present in everything we can imagine.

And bearing in mind that works to coexist with the semi-arid region such as plate cisterns, among others, are extremely important for the survival of family farmers, given the difficulties of producing food in such inhospitable regions. Thus, applied mathematics emerges around all these activities, making this reality easier and more enjoyable to experience. It is in this context that we mention the study of ethnomathematics in the words of the late Ubiratan.

> Ethnomathematics is not just the study of "the mathematics of various ethnic groups". To create the word *Ethnomathematics,* I used the roots *tica, matema and etno* to mean that there are various ways, techniques, skills (*tica*) to explain, understand, deal with and live with (*matema*) different natural and socio-economic contexts of reality (*etno*).
>
> (D'Ambròsio, 1997, p. 111 and 112)

As a result, a number of questions were asked in order to come up with the topic. Why do farmers build buildings or deal directly with practices that involve a lot of mathematical knowledge, without ever having attended school? How did they acquire this knowledge and become skilled in these practices involved in their daily tasks? Well, how can that be? People who are almost or totally illiterate can develop mathematical skills that amaze us!

To do this, we had to study authors in this area such as Ubiratan D'Ambròsio and Paulo Freire, since the bibliographical research related to the study of this knowledge is directly linked to Ethnomathematics and literacy knowledge. According to D'Ambròsio (2005), mathematics lessons should be based on mathematical knowledge that is brought into the school from outside. This knowledge should be developed from the student's own life experience.

It is with this in mind that we intend to explore this work. To show those who teach mathematics that it is possible to develop techniques and tools aimed at integrating mathematical theory with everyday practice. This makes it possible to develop and learn mathematics not only in the classroom, but also in everyday life, in a constructive process that integrates this knowledge with practice.

The aim of this research is to investigate and analyze through ethnomathematics how farmers and their closest relatives can learn mathematics during the construction of a slab cistern.

Our hypothesis is that the practice of building the cistern will contribute to the students and those involved with the mathematical knowledge present in this construction. This activity also appropriates elements of mathematics that are developed in the classroom.

On the other hand, this work has a structure of chapters that are organized as follows: in chapter 1, we have the introduction that addresses the initial idea of the work that follows with the interest in the theme and the problematic of the research and ends with the structure of the work.

Chapter 2 presents the theoretical and methodological foundation of the research, which is based on the ethnomathematical approach. Also in this chapter, we have the qualitative research, the participants in this research, the records and data collection and the description and objects of the questions in the questionnaire.

In Chapter 3, we present the analysis of the intervention in the community, with a little background on the Barra do Japi community, the justifications, the results of the interviews and the analysis of the socio-economic data of the interviewees. This section also includes a more detailed analysis of the construction process of a slab cistern, from the marking process to the covering of the cistern, all from the point of view of ethnomathematics, so that the procedures adopted and the research methodology used can lead us to the formation of knowledge in an interactional educational context from the point of view of non-formal education.

Finally, in Chapter 4, we present the step-by-step process of learning the mathematics involved in building a plate cistern. With this, we tried to work on mathematical content such as: calculating area and volume involving plane and solid geometry, radius length, diameter and circumference, as well as the costs of building a cement slab cistern with a capacity of 16.33m .^{3}

Finally, in the final considerations of the work, we tried to approach the analysis achieved from an educational point of view, aiming at a more detailed analysis of whether it is possible for farmers and their close relatives to learn mathematics in the course of their daily practices.

2 THEORETICAL AND METHODOLOGICAL BACKGROUND

Through this chapter, we intend to show our methodological proposal that guided the field research and analysis. And bearing in mind that when we work on the mathematics curriculum we have to consider the fact that we are doing mathematics education by integrating knowledge with practice. Ubiratan reinforces this in his words:

In moving the discussion towards the possibility of doing Education through Mathematics during classes, I understand that the Mathematics curriculum also collaborates in developing the ability to mathematize real situations, codifying them appropriately, in such a way as to allow the use of known techniques and results in another context.

Ubiratan D'Ambròsio

2.1 Research participants

With regard to the participants in this research, we tried to focus our efforts on family farmers in a particular rural community called Sitio Barra do Japi, located on the border of the rural territory between the municipalities of Santa Cruz/RN and Japi/RN. We noticed that there are people there who, when working on construction sites or in everyday activities, use mathematical knowledge without hardly realizing it. So we have a situation that already exists, but we are going to analyze and investigate it. In this way, 15 farmers were interviewed in their homes, where they had or had not been directly involved in activities related to a slab cistern.

2.2 Justifications

Given the need to carry out an intervention in order to initially address a previous situation regarding the issue we want to work on with the community and the objective we want to achieve, we had to draw up a qualitative questionnaire in order to first understand to what level we should work with the community regarding the problem raised. We now have the important task of analyzing these responses.

Thus, according to Ludke and André (1986), when analyzing qualitative research, we must work with the research material throughout the process. What's more:

The task of analysis involves first organizing all the material, dividing it into parts, relating these parts and trying to identify relevant trends and patterns in it. In a second step, these trends and patterns are realized, looking for relationships and interferences at a higher level of abstraction.

(Ludke and André, 1986, p.45).

2.3 The ethnomathematical approach to methodology

When thinking about the research, we focused on Ethnomathematics, which allows us to differentiate it from research into teacher training and non-formal education. Ethnomathematics forms the main axis of the research that guides this work. It also provides us with an enriching basis for arriving at our hypothesis. We cannot dissociate education and culture, with a view to teaching and learning mathematics.

I see education as a strategy to stimulate individual and collective development, generated by cultural groups in order to maintain themselves as such and to advance in satisfying survival and transcendence needs. Consequently, Mathematics and Education are contextualized and totally interdependent strategies.

Ubiratan D'Ambrósio

Legislation also supports us in this regard. The

The Guidelines and Bases state more specifically in Article 28 that:

Art. 28: In the provision of basic education for the rural population, education systems shall make the necessary adjustments to suit the peculiarities of rural life and of each region, in particular:

I - curriculum content and methodologies appropriate to the real needs and interests of rural students;

II - proper school organization, including adapting the school calendar to the phases of the agricultural cycle and climatic conditions;

III - suitability for the nature of work in rural areas

(BRASILIA, 2007).

2.4 Qualitative research

Our work proposal involves qualitative research, since we are focusing on curriculum and teacher training in the process of non-formal education. This is therefore the best methodology to suit our needs. Thus, according to Ambrózio, *"This type of research is typical of field research, where the theoretical framework, which results from the researcher's philosophy, is intrinsic to the process". (p.102-103).*

Therefore, with the guiding focus of the research, and being a qualitative research, we structured the following script:

a) Elaboration of the questions to be investigated in the field intervention a priori;

b) Identification of the location, subjects and objects that will make up the research;

c) Strategy and definition of data collection and analysis;

d) Data analysis and refinement of the questions posed.

2.5 Research records and data collection

This was the stage at which we first started working with the community. It was a very important phase, since it involved getting to know the actual people we were going to work with. We had to check the details we were going to face in terms of access logistics, intervention time, suitable materials, work environment and space, as well as aspects related to the application of the questionnaire.

After choosing the location and target audience for the research, we began our fieldwork. This took place on January 16 and 23, 2017, more specifically on two Mondays. But first we had to draw up a whole approach strategy. From the first contacts to the questionnaire application phase and verification of the stages in the construction of a slab cistern.

This was very gratifying for me and somewhat reassuring, as this is the same audience that I have been working with for some time now, through EMATER-RN. I'm part of their technical Rural Extension team.

The questionnaire was applied directly to their homes through one-to-one visits. First we explained the purpose of the questionnaire and then we checked the details question by question and recorded everything in the context of the reality experienced at the time.

2.6 Description and objectives of the survey questions

The general objective is to investigate and analyze through ethnomathematics how farmers and their close relatives can learn mathematics during the construction of a slab cistern. It also has specific objectives, which are directed towards the questions in the questionnaire, which has a total of 14 questions.

The questions were selected with themes that are part of the daily lives of the participants in this research. In order to obtain results that did not differ so much from their reality. It was with this in mind that we drew up the script and objectives for the following questions:

- Question 01 - How is your house built?

Objective: To find out whether rammed earth houses still exist, as they are already banned under current legislation.

- Question 02 - How is the water supply provided to your home? **Objective: To** find out whether or not the family is assisted by any public water policy, especially the policy of the plate cistern or water tanker.

- Question 03 - Do you have a slab cistern?

Objective: To identify whether this public policy project that brings this type of construction to rural people has reached 100% or is still inefficient.

- Question 04 - Where is the water stored?

Objective: To identify if the family does not have a plate cistern, where they store it and if they store any water for consumption.

- Question 05 - Do you have an antiseptic tank in your house?

Objective: In this case we intend to analyze whether there are still families who do not treat their sewage properly.

- Question 06 - Do you reuse water for bathing and domestic use? **Objective: To** identify whether there are any families who think about environmental sustainability.

- Question 07 - Up to what grade or year did you study?

Objective: Here we intend to relate schooling to the activity of building the slab cistern and other existing activities.

- Question 08 - Can you sign your name and/or read and write?

Objective: To identify whether there are still families who are not literate and to compare this with the data from the previous question in order to assess the level of efficiency of their studies.

- Question 09 - Which of the basic mathematical operations have you mastered the most?

Objective: To find out if they have mastered mathematics more when it comes to calculating bills in everyday activities.

- Question 10 - Do you know what geometry is?

Objective: The aim is to identify whether the target group has any knowledge of geometry. In order to associate it with the work on the plate cistern.

- Question 11 - What is your profession?

Objective: To associate the activity of building the cistern with the beneficiary's profession. In order to find out whether the beneficiary uses the skills of his profession when building the cistern.

- Question 12 - Do you use mathematics for any day-to-day activities? **Objective: The aim** is to identify whether mathematical ability comes from everyday activities.

- Question 13 - How is what is produced in the community marketed?

Objective: To find out whether marketing is carried out with the right degree of entrepreneurial logistics.

- Question 14 - What is your family's monthly income?

Objective: The aim is to identify the socio-economic conditions of the target group. In order to evaluate the effect of the numerous public policies installed with these families.

3 ANALYSIS OF THE COMMUNITY SURVEY

In this chapter we will present a qualitative analysis using a questionnaire that was applied to family farmers in the Barra do Japi community during the intervention that was carried out. We intend to present tables and graphs generated from the results of this questionnaire with the farmers. These include summaries of questions such as: level of education, social class, professional activity, how did you learn mathematics, since when? In addition to the question related to the water situation in the slab cistern.

3.1 The Barra do Japi community

The community is a traditional rural community called Sitio Barra do Japi, located in the Trairi territory, in the Agreste mesoregion and Borborema Potiguar microregion of the state of Rio Grande do Norte (IBGE - 2008). It is located on the border between the municipalities of Japi/RN and Santa Cruz/RN, about 20 km from the city of Santa Cruz and around 10 km from the city of Japi/RN, accessed via the RN 092. It is a community made up of family farmers who have been farming and ranching since the time of the first inhabitants. Its residents live basically from what they produce and also from the rural retirement salaries of the elderly. With a Catholic majority and traditional customs, they enjoy a peaceful life with all the support in the areas of education and health provided by the municipal government of Japi and Santa Cruz.

3.2 Analysis of results

Here we will present an analysis of the questions asked during the field intervention. The aim is to draw an analogy between what we wanted to achieve in the light of the methodological foundation and the reality we found in the locality.

In order to make use of this task, we tabulated and graphed the results of the field research intervention. The results of the analysis in Table 1 show that there is a predominance of male interviewees, 80% of whom were male and only 20% female.

It is also noticeable that among the participation of these people in everyday activities in the countryside, there is still a very strong predominance of men. This goes against national guidelines which call for greater female participation in social inclusion tasks in the face of these public policy projects.

Chart 1: Frequency and percentage of interviewees' gender.

Gender	Frequency (fi)	fr (%)
Male	12	80
Female	03	20
Total	**15**	**100**

Source: *Prepared by the author.*

On the other hand, with regard to the age criteria of the interviewees, we had a very interesting frequency. The participation of people over the age of 60 represented 60%, as shown in Table 2 and Graph 1 below:

Table 2: Frequency and percentage of interviewees' ages.

Age (years)	Frequency (fi)	fr (%)
Up to 40	03	20
Between 40 and 60	03	20
Over 60	09	60
Total	**15**	**100**

Source: *Prepared by the author.*

Graph 1: Frequency of interviewees' ages.

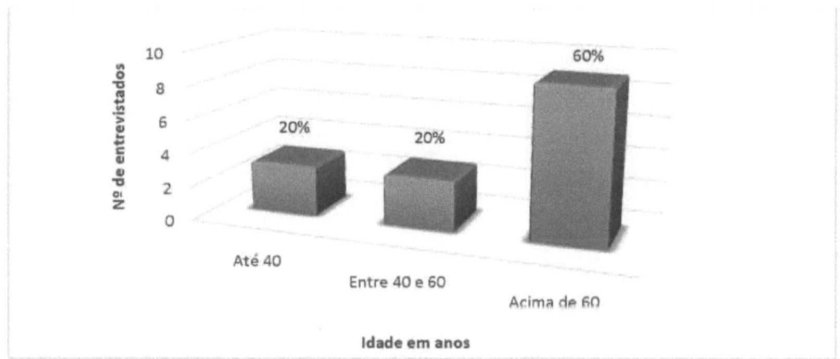

Source: *Prepared by the author.*

After this initial analysis of the genders and ages of the interviewees, we will begin our analysis of the questions in the questionnaire, starting with Question 01.

- ***Question 01 - How is your house built?***

This question revealed that 100% of those interviewed live in a masonry house. This leaves us to conclude that nowadays it is very difficult for someone to still live in a rammed earth house, as we intended to analyze in this item.

- ***Question02 - How is the water supplied to your home?***

In this question we wanted to know if there was a public water policy in the community. However, the answers show that there is indeed such a policy. According to the data in Table 3 and Graph 2, 40% of the interviewees receive their water from a fountain, which in turn is connected to the water main. And one person did not have this supply policy.

Table 3: Frequency and percentage in relation to the type of water supply.

Water supply (x)ᵢ	No. of people (fi)	fr (%)
Pipeline only	01	6,66
Fountain only	06	40
Water tanker only	02	13,33
Water main and fountain	02	13,33
Fountain and water cart	02	13,33
Water tanker and others (Well)	01	6,66
Fountain and others (well)	01	6,66
Total	**15**	**100**

Source: *Prepared by the author.*

Graph 2: Frequency of respondents' water supply.

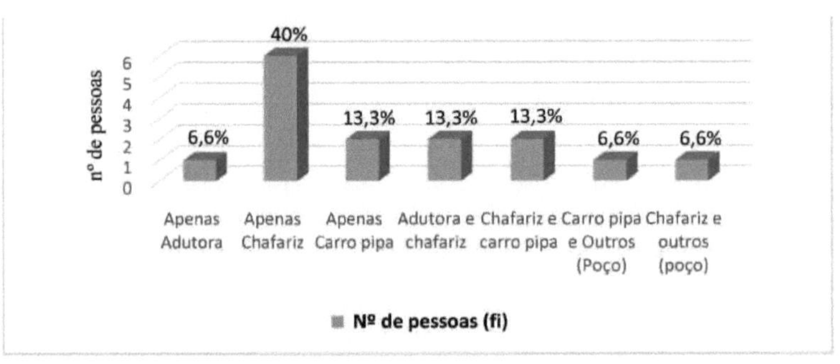

Source: *Prepared by the author.*

- ***Question 03 - Do you have a slab cistern?***

Proceeding with the analysis, we can see that in this item among the interviewees we obtained a total of 12 people, representing 80% of those interviewed who have a plate cistern in their home. As a result, we saw that this public policy of building plate cisterns in rural areas is very strong. However, there are still 3 families among those interviewed who do not have a cistern in their home, as shown in table 4 below.

Table 4: Frequency and percentage with regard to the slab cistern.

Existence of a slab cistern in your home (x)$_i$	No. of people (f_i)	fr (%)
Yes	12	80
No	03	20
Total	**15**	**100**

Source: *Prepared by the author.*

- ***Question04 - Where is the water stored?***

For this question, we unanimously found that all the interviewees store water in slab cisterns for those who have them and in masonry cisterns or water tanks.

- ***Question05 - Do you have an antiseptic tank in your house?***

As with the previous question, we also got the response that all the interviewees have cesspits in their homes. This is very good, because it shows that there is treatment of domestic waste.

- ***Question06 - Do you reuse bath water and household utilities?***

Of the 15 interviewees, only 3 said that they reused water for bathing and domestic use. As this item was intended to verify whether there was this concern with the reuse of bathing water.

Well, we found that only 20% of those interviewed reuse this water. This reuse is still precarious. Ideally, this water should be decanted and filtered before being reused. We found that they only direct the water through small open ditches and use it to water small orchards in their backyards.

- ***Question 07 - Up to what grade or year did you study?***

For question 7 of the questionnaire, we wanted to investigate and compare the level of education of the interviewees in relation to the activity of the slab cistern and other activities. On the basis of table 5 combined with the data in graph 3, we found that the predominant results were people with between 0 and 2 years of schooling, i.e. people with up to 2ª years of primary schooling were more likely to be involved in these activities.

However, we also saw a small increase in these figures for people with 8 to 10 years of schooling, as shown in table 5 below.

Table 5: Frequency and percentage in relation to level of education.

Years of study (xi)	**No. of people** (fi)	(fac)	**(PM)**	**(fi.PM)**	**fr (%)**
0 |- 2	06	06	01	06	40
2 |- 4	02	10	03	06	13,3
4 |- 6	02	11	05	10	13,3
6 |- 8	01	14	07	07	6,7
8 |- 10	03	14	09	27	20
10 |- 12	01	15	11	11	6,7
Total	**15**			**67**	**100**

Source: *Prepared by the author.*

Graph 3: *Frequency of time spent studying by interviewees.*

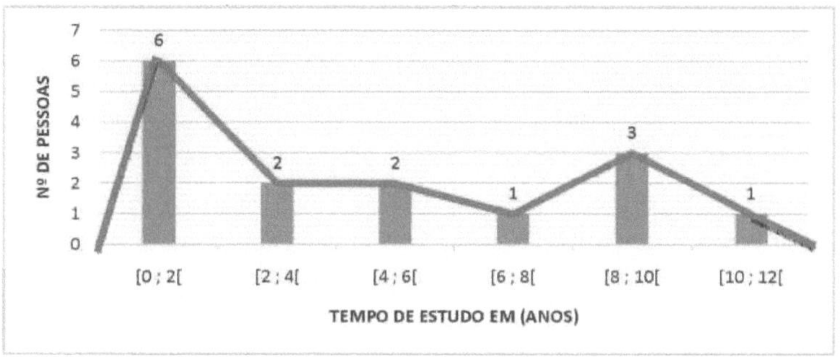

Source: *Prepared by the author.*

These two phenomena can be clearly seen in the histogram graph below, where we see the normality curve showing a stronger symmetry for people with 0 to 2 years of schooling and a slightly weaker symmetry for those with 8 to 10 years of schooling. We therefore conclude that the strongest symmetry concerns older people, who have had almost no opportunity to study. The other symmetry shows that young people are becoming increasingly involved in these activities. This was only possible by comparing the data in table 5 with the data in table 2.

- Question08 - Can you sign your name and/or read and write?

In order to compare the data from these answers with the previous question, we see that table 6 reveals a phenomenon related to people who only sign their name. These people claim to be literate or to have studied up to grade 2ª . However, 40% of those interviewed tried to become literate, but only managed to learn to sign their own name. This is what we call functional illiterates.

Table 6: Frequency and percentage in relation to literacy level.

Can you sign your name and/or read and write?	No. of people (fi)	fr (%)
Just sign	06	40
Don't sign	01	6,6
Can only read	02	13,3
Can read and write	06	40
Total	**15**	**100**

Source: *Prepared by the author.*

Graph 4: Frequency of interviewees' level of literacy.

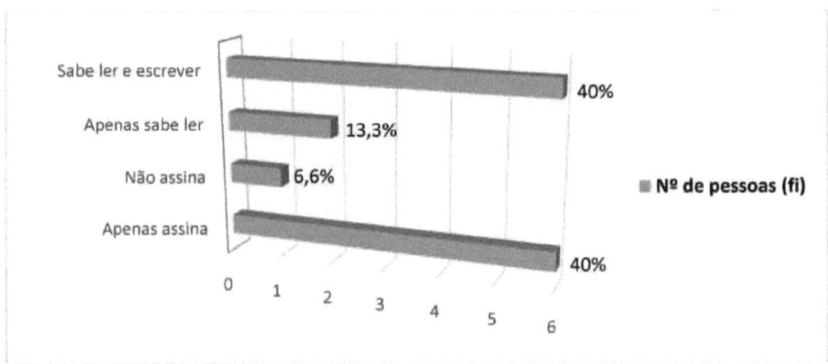

Source: *Prepared by the author.*

Only one of the interviewees doesn't know how to sign, i.e. is completely illiterate. The rest are literate. With 40% of those interviewed being able to read and write. This is a growing phenomenon that we already noted in the previous question about the increasing inclusion of young people in these public policies.

Therefore, the orientation at the national level is to include young people more and more in these policies.

- ***Question 09 - Which of the basic mathematical operations have you mastered***

the most?

Here we saw that both interviewees said they had mastered the operations of adding and subtracting. However, only 26% said they knew how to divide and multiply.

These answers revealed what we wanted to investigate. Their knowledge of adding and subtracting is directly related to their everyday activities.

- ***Question 10 - Do you know what geometry is?***

Of those interviewed, 40%, or 6 people, said they knew about geometry. But in a very vague way. The rest said they didn't know. This shows us that there is still a public that is involved in cistern-building activities, but does not yet know about geometry.

- ***Question - What is your job?***

In this item, all of them revealed that they are farmers. This shows us that the farming profession is closely related to basic mathematics in terms of soil preparation, production cycles, harvests, storage and marketing. That's why they have excellent basic math skills. Especially when it comes to adding and subtracting.

- ***Question 2 - Do you use mathematics for any everyday activities?***

In relation to question 12, we found that 4 out of the 15 people interviewed said that they use mathematics to calculate household bills. What stands out, given that the profession of the interviewees was that of farmer, was the fact that only 9 of the interviewees said that they use mathematics in agricultural marketing.

- ***Question 3 - How is what is produced in the community marketed?***

The highlight here was that 13 of the interviewees said that they sold their agricultural products in the town's own shops. The rest went to local or regional shops.

- ***Question 4 - What is your family's monthly income?***

As was to be expected in view of the various previous results in the graphs and tables. Here we have confirmation that we are working with a predominantly elderly audience over the age of 55. Table 7 and graph 5 clearly show the analysis we are dealing with.

Table 7: Frequency and percentage in relation to family income.

Family Income (R$)	No. of people (fi)	fr (%)
Up to 1 S.M.	03	20
Between 1 and 1.5 S.M.	04	26,6
Above 2 S.M.	08	53,3
Total	**15**	**100**

Source: *Prepared by the author.*

Graph 5: Frequency of interviewees' family income.

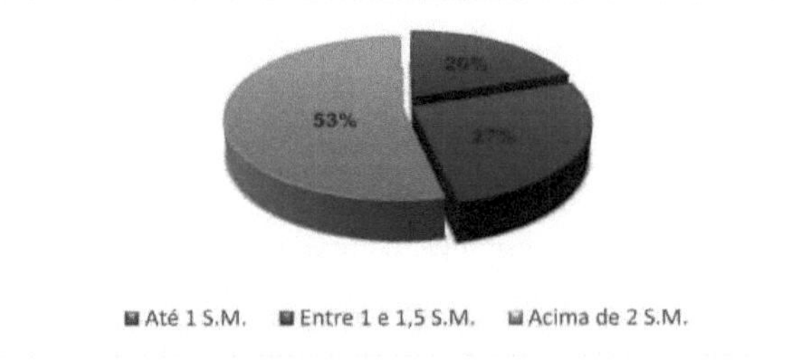

Source: *Prepared by the author.*

When we look at table 7 and graph 5, we see that the fact that these people are all farmers and that women retire at the age of 55 and men at the age of 60. It can be deduced that among the 53.3% who said they had more than two minimum salaries, this is because both the woman and her husband are already retired on two minimum salaries plus what they produce from their agricultural income.

3.3 Analysis of the construction process of the slab cistern

In order to build a slab cistern, the owner must have the following materials at his disposal in addition to the building materials: sand, cement, iron, gravel, forms for making the slabs and roof beams. You also need to hire a specialized workforce of bricklayers and servants. And most of the time it is the farmer himself who carries out this task of mason and servant, because he has this skill in his family.

If you need to build a slab cistern, in addition to the materials mentioned above, you also need a good location for it. It needs to be close to your home, at a distance of about 3 or 4 meters, so that it is not far from the supply system. The system will be fixed directly to the roof of your house and will work through gutters installed to catch the water that falls from the rain.

3.3.1 The slab cistern

The slab cistern project emerged in 2000 in the northeast of Brazil in response to the urgent need to store water in the region most affected by the great droughts, the Brazilian semi-arid region. It was then that a group of researchers from an organization called ASA (Articulaçao Semiàrido) created the plate cistern project. They began to implement it in the Northeast through a government program called 1 Million Cisterns.

The slab cistern, as the name implies, is made of cement slabs, its base is circular in shape and its volume is cylindrical with a cone-shaped lid. Today it has various capacities in terms of volume. More precisely, we'll be working with a cistern with a capacity of 16.33 m^3 . This cistern is basically used to store water in the rainy season through gutters installed on the roofs of houses and in other times of drought through the Army's water tanker program.

In turn, the slabs that make up the walls of this cistern are rectangular in shape. This allows us to work on geometry by calculating the area of the side walls so that we know how many slabs we need to build it. The base is also circular and this allows us to calculate the diameter, radius and area of the base to find the ideal height so that we have a capacity of 16.33 m3 of water.

In short, the slab cistern will help us do a good job of learning the mathematics behind its construction. To build it, you need the following materials as well as a good location: sand, cement, iron, gravel and wooden molds to make the slabs and beams.

3.3.2 Choosing the right land and marking out the cistern

The land must be flat and close to the residence of the farmer who will build it, as shown in image 1 below. As for its marking, this will be in a circular shape and with a diameter of 5 meters for a water volume of 16.33 thousand liters of water. This volume is enough for a family of 4 to meet their water needs for approximately 3.5 months.

Taking into account that it is assumed that each person in a family spends at least 40 liters of water a day to meet basic hygiene and survival needs in terms of bathing, drinking, cooking and washing dishes.

This is why we need to choose a good plot of land so that the cistern is close to the family's home, so that they can always keep it well maintained and with enough water for their maintenance.

Image 1: Choosing the site for the slab cistern

Source: *Own collection*

This marking must be made with a diameter of 5 meters, as the actual diameter of the cistern is 3.40 meters by 1.8 meters and the depth of the foundation is 1.2 meters, i.e. the cistern is 1.2 meters underground to better support its lateral structure.

The extra space in the 0.8 meter marking is due to the fact that the professionals who are going to build need a comfortable space to work inside the excavated hole. This marking is done using a string tied to a picket in the center of the circle with another picket at the other end. In order to make the right mark, the bricklayer will need to make a 360° turn, as shown in figure 02 below.

Image 2: Marking the hole in the plaster cistern

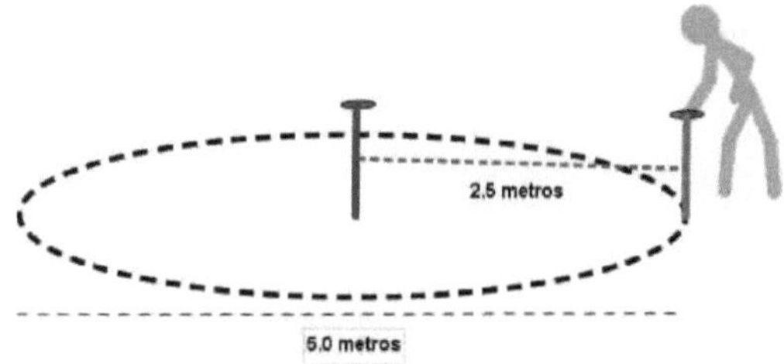

Source: *Own collection*

3.3.3 Leveling the floor and building the wall

The floor is leveled when the circular hole is dug. This hole is dug with half a meter to spare on each side so that the bricklayer can work loosely to build the wall. Once the excavation is complete, the actual leveling of the floor begins.

This will be done by filling in with a concrete subfloor and an iron mesh. The leveling begins by constantly checking with a leveling hose. This is used to determine that the floor is actually level by means of the water inside it.

Image 3: Leveling the floor

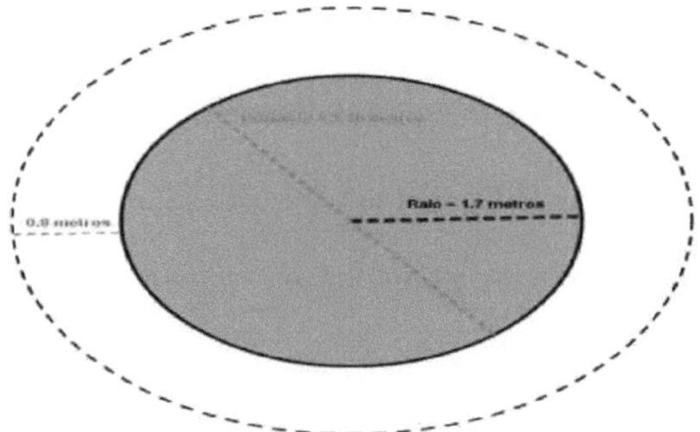

Source: *Own collection*

The side wall is built with rectangular cement blocks measuring 50 cm wide by 60 cm long and 3 cm thick. They are made using wooden forms. Once they are ready, they are laid one by one, making the joints as shown in image 4 below:

Image 4: Construction of the cistern wall

Source: *Own collection*

3.3.4 Roof construction

The use of this type of cone shape allows the roof structure to be supported more efficiently, since the cone shape is able to support more weight than the flat roof shape. The slabs are also made using a form in the shape of an isosceles trapezoid whose measurements are: base larger, 50 cm, base smaller, 8 cm and sides 1.63 cm. This form has two dividers that allow three slabs to be made at once. See image 5 below:

Image 5: Cover of the slab cistern

Source: *Antônio Conselheiro Institute website*

4 MATHEMATICS IN THE CONSTRUCTION OF A CISTERN PLATES

Learning mathematics doesn't just take place in the classroom, but also outside of it and in a dynamic way that allows theory to interact with practice. It is with this in mind that we agree with D'Ambrózio when he states that:

Using everyday shopping to teach mathematics reveals practices learned outside the school environment, a true ethnomathematics of commerce. An important component of ethnomathematics is to enable a critical view of reality using mathematical tools. Comparative analysis of prices, accounts and budgets provides excellent teaching material.

(D'AMBROSIO, 2001, p.23).

4.1 Measuring the length of the circle, radius and diameter.

At this stage, we've tried to introduce the concepts of circumference, radius and diameter in order to provide a better basis for our proposal. So let's get to the concepts. First of all, we need to understand that a circle and a circumference are similar but different. A circle is any internal circular part that is bounded by the circumference. A circle, on the other hand, is just the part that limits the inner circular part from the outer non-circular part.

- Definition of circumference:

*A **circle** is a set of points belonging to the plane which, given a fixed point C, have the same distance to point C. In other words, given a distance "r" and a fixed point C, any point A which has a distance from A to C equal to r is a point on the circle.*

Now we can delve into the step-by-step process of how to make the markings to start building a slab cistern. First of all, we start by choosing a plot of land. It should be flat and close to the house. Then we choose a point that will be the center of the circle that will be the base of the cistern.

To mark the length of the circle. In this case, you bury a picket in the center of the circle and with a string tied to the picket and stretched out to measure the radius of the circle, you mark the circumference by starting a circular path with another picket at the other end of the string until you form a 360° turn and close the circle.

The result of this initial work is the formation of a circular base for the plate cistern. Where we can also learn: (basic knowledge) show images of the geometric figures you are going to use.

- Diameter: $D = 2 \cdot r$
- Circumference length: $C = 2 \cdot \pi \cdot r$

Pi (π) is an irrational number, which we will consider rounding up to 3.14 in our problems. To better show and illustrate what we have just detailed, we then have the images below illustrating the comparison between the real photo and the Geogebra image illustrating the geometric result.

Image 6: Studying the circumference with the plate cistern

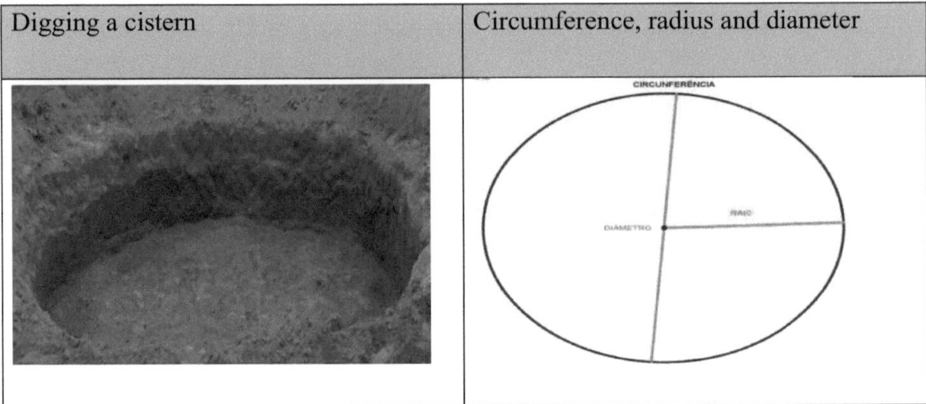

Source: *Amargosa Diocesan Caritas website*

4.2 Measuring the area of the circular base and the area of the side wall

To do this, let's first understand the flat geometric figures that are directly linked to these calculations. The figures in question are: the circle, the rectangle and the triangle. After understanding the main figures involved in the cistern. Now we can actually start calculating the area of the base and the area of the wall. So we have:

Calculating the area of the circular base of the plate cistern:

Area of the base: $A_b = \pi \cdot r^2$, where$\pi \cong 3.14$ er is the radius of the circle.

Image 7: Calculating the area of the circular base of the cistern

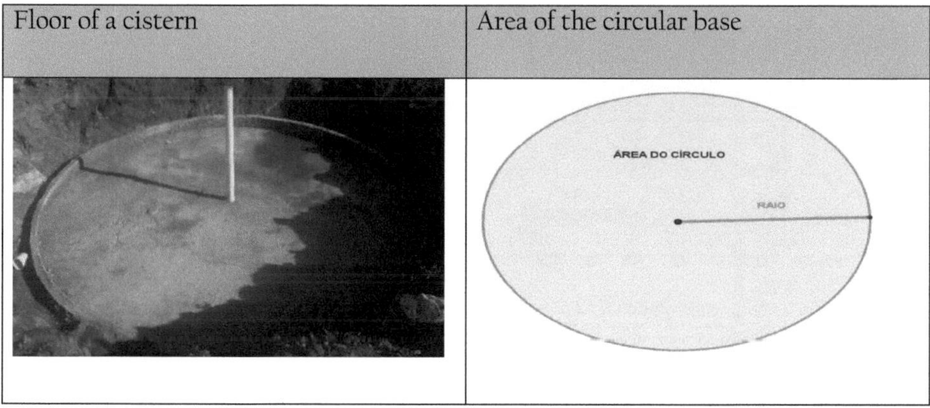

Source: *Own collection*

Example 01:

Calculate the area of the base of a cistern with a radius of 1.7 m.

Solution:

$$A_b = \pi \cdot (1,7)^2 \cong 3,14 \cdot 2,89 \cong 9,07\ m^2$$

- Calculating the area of the side wall:

We can see that the side surface is a rectangle with base $2 \cdot \pi \cdot r$ and height h. Thus, the area of the side surface will be given by:

Image 8: *Planning the cistern*

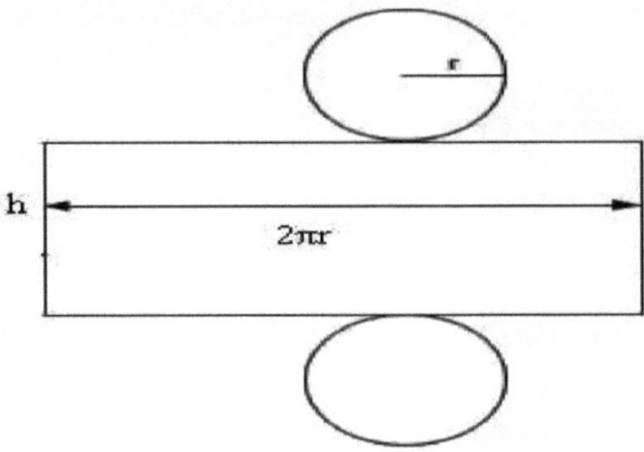

Source: *Own collection*

Area of the side: $A_l = 2 \cdot \pi \cdot r \cdot h$, where h *is* the height of the cylinder and r is the radius of the base.

The total area of the cylinder is obtained by adding the area of the two bases to the lateral area. In this way, we will have:

Total area: $A_t = A_l + 2 \cdot A_b$ (eq. 1)

Lateral area: $A_l = 2 \cdot \pi \cdot r \cdot h$ (eq. 2)

Area of the base: $A_b = \pi \cdot r^2$ (eq. 3)

It follows that:

$$A_t = 2 \cdot \pi \cdot r \cdot h + 2 \cdot \pi \cdot r^2$$

Or

$$A_t = 2 \cdot \pi \cdot r \cdot (h + r)$$

Example 02:

Calculate the area of the side wall of a cistern that is 1.8 m high and has a radius of 1.7 m from the base.

Solution:

$$A_l = 2 \cdot \pi \cdot r \cdot h \cong 2 \cdot 3{,}14 \cdot 1{,}7 \cdot 1{,}8 \cong 19{,}21\ m^2$$

On the other hand, we can see that the area of the wall allows us to calculate how many slabs we need, depending on their size, to form the entire wall of the cistern. Thus, after planning the area of the entire wall, we have a large geometric figure, the rectangle.

We also know that one plate will also form a rectangle. Therefore, easily dividing the area of the entire wall by the area of one plate gives us the number of plates we will need to form the entire wall of the cistern.

Image 9: Planning the area of the cistern's side wall

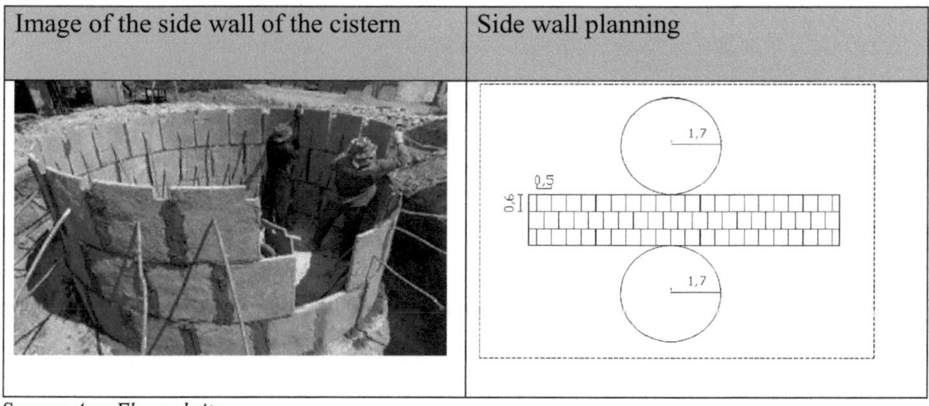

Source: *AgroFlor website*

4.3 Measuring the volume of the slab cistern

To do this, we need to understand what kind of geometric solid figure a plate cistern represents. We know that the part of the cistern that we take as volume is only the part that will be completely filled with water. So for the purpose of calculating the volume of water, we only consider the cylindrical part and not the roof, which *already* represents another geometric figure that we'll talk about later. So to understand what a cylinder is, let's look at the concept. According to Dolce and Pompeo (2005) a cylinder is:

- Cylinder definition:

*The **cylinder** is the reunion of the part of the unlimited circular cylinder between the pianos of its parallel circular sections and distinct in relation to these sections.*

Image 10: Cylindrical volume of the cistern

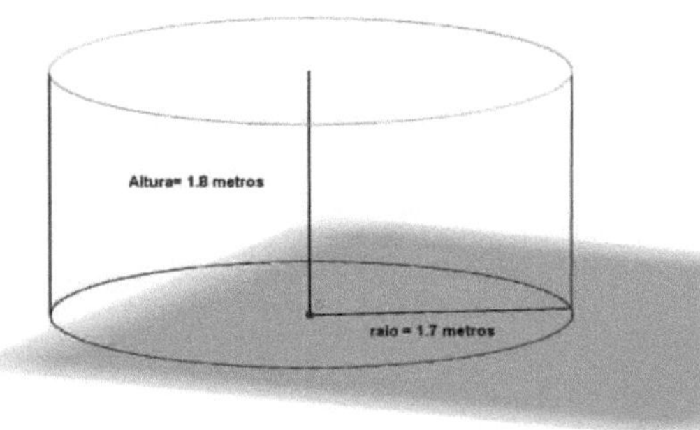

Source: *Own collection*

The volume of a cylinder, according to Cavalieri's principle, is obtained in the same way as the volume of a prism. We can therefore say that the volume of a cylinder is equal to the product of the area of the base and the height, or

$$V = A_b \cdot h = \pi \cdot r^2 \cdot h \quad \text{(eq.4)}$$

It follows that,

$A_h = \pi \cdot r^2$, according to eq. 3

Note that image 11 shows a comparison between a photo of a real cistern and an illustration drawn in cylindrical form.

Image 11: Calculating the volume of the slab cistern

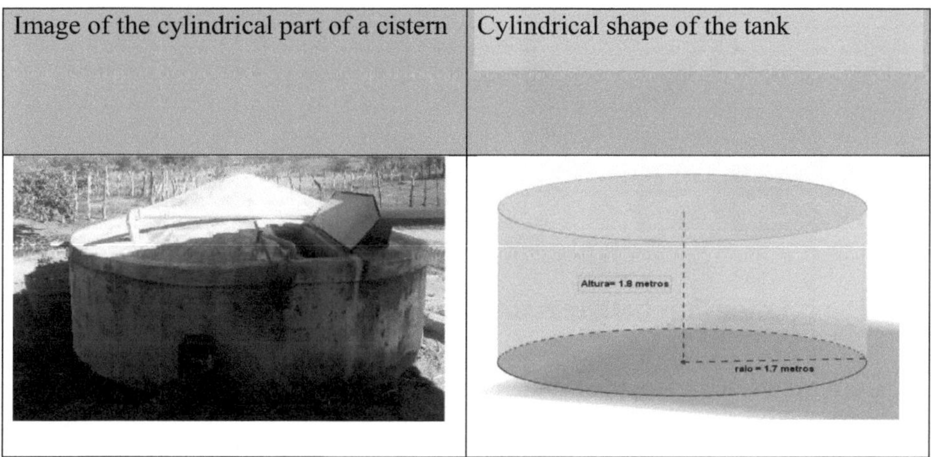

Source: *Own collection*

Example 03:

Calculate the volume of a cistern that is 1.8 m high and has a base radius of 1.7 m.

Solution:

$$V = 3{,}14 \cdot (1{,}7)^2 \cdot 1{,}8 = 3{,}14 \cdot 2{,}89 \cdot 1{,}8 = 16{,}33\ m^3$$

4.4 The cone-shaped cover

The roof is conical in shape, and in its structure we can see that the beams are distributed over the wall to give better support to the trapezoidal plates, i.e. plates in the shape of trapezoids. Thus, the whole roof forms a conical geometric figure. We can easily see from image 12 that a cone has the following parts: base, side face, vertex, radius and height.

Image 12: Cone study with the plate cistern

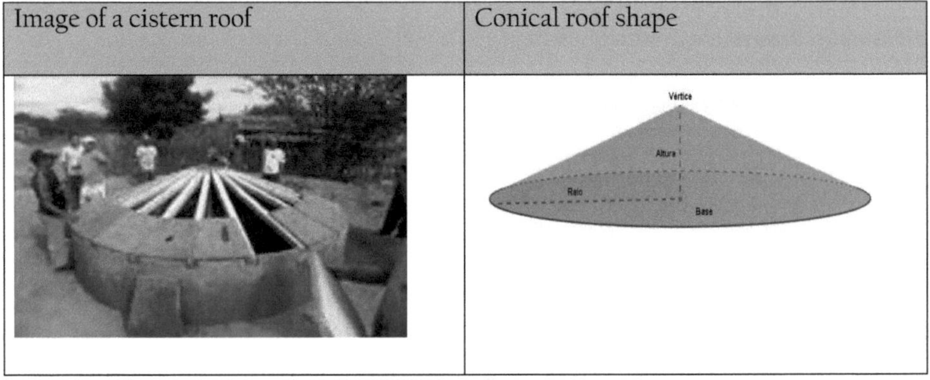

Source: *https://www.youtube.com/watch?v=0GNu2Rr0-dc*

4.5 The relationship between volume and rainfall

So how do you calculate rainfall storage? It works by calculating the ratio of 1mm of rain to Im^2 of the roof of the house, which is equivalent to 1 liter of water. In other words, if I want to complete the cistern's capacity of 16,000 liters, it has to rain how many mm of rain? And for a family of 6 people using about 40 liters of water a day, what would the calculation be? And for how many days would this cistern be able to make up for this family's water deficiencies? These are the questions and others related to mathematics that the construction of this cistern can teach us.

Image 13: Cistern supply system

Source: *Own collection*

4.5.1 Problem 01 - Estimate the total cost of building a cistern made of cement slabs with dimensions of 1.8 m high and 3.4 m in diameter. Consider the cost of R$ 200.00/m^3 for mortar based on cement and sand formulated in the type 1: 4.5 (4.5 cans of sand are needed for 1 can of cement) and the cost of R$ 240.00/m3 for concrete formulated in cement, sand and gravel in the type 1: 2: 2 (2 cans of sand and 2 cans of gravel are needed for 1 can of cement). These prices are based on the cost of materials found on the market. And that the cost of labor is approximately Æ$ 1100.00.

a) Calculate the volumetric quantity and the cost of making the slabs for the side wall, considering that 63 rectangular cement slabs with dimensions of 0.50m x 0.60m x 0.03m will be needed; (Justify why 63 slabs and not 64);

b) Calculate the volumetric quantity and cost of the lid slabs, considering that 19 cement slabs with an area of 0.41m^2 and a thickness of 0.03m each will be needed to build the cistern lid.

c) Calculate the cost of making 19 cement beams measuring 1.7m *x 0.06m x* 0.08m to support the slabs on the lid.

d) Calculate the volume and concrete cost of the cistern floor, considering a thickness of 7 cm;

e) Now calculate the volumetric quantity and cost of the mortar needed to build the internal plaster on the side wall of the cistern, knowing that the thickness of the plaster is 0.02 m;

f) And finally add it all up to get the total cost of the cistern.

Solution (item a):

First we calculate the cost of making the side wall panels. So, according to *equation 2* **on page 33**, we have a wall area of 19.33 m^2 . We then realize that in order to get the number of slabs we had to divide the side area by the area of one slab, which is 0.30 m2. In this case, the result is 64.4 slabs. However, you will only need 63 slabs to build the side wall, as you will need a small gap of about 2 cm between the slabs to be grouted with cement.

Now, to find out the cost, we have to calculate the volume of each side wall plate (V_{pl}). This follows,

$$V_{p1} = 0{,}50 \cdot 0{,}60 \cdot 0{,}03 = 0{,}009\, m^3$$

In other words, it takes 0.009m^3 of mortar to build each rectangular slab. Therefore, the price of one slab is 0.009 - 200.00 = Æ$ 1.80 and the total cost of the rectangular slabs is

$$63 \cdot 1{,}80 = R\$\ 113{,}40.$$

Solution (item b):

We need to calculate the volume of each lid plate (V_{p2}). Therefore,

$$V_{p2} = 0{,}41 \cdot 0{,}03 = 0{,}012\ m^3$$

Thus, 0.012m3 of mortar is needed to build each cover plate. Therefore, the cost of one cover plate is 0.012 - 200.00 = Æ$ 2.40 and the total cost of the 19 plates will be

$$19 \cdot 2{,}40 = R\$\ 45{,}60$$

Solution (item c):

Now, to calculate the cost of making the 19 beams, we will calculate the volume of one beam (V). Thus,

$$V_v = 1{,}7 \cdot 0{,}08 \cdot 0{,}06 = 0{,}008\ m^3.$$

[3]Therefore, the cost of concrete to build 1 beam is 0.008 - 240.00 = R$ 1.92 and the cost of iron to build it is 1.7 - 3.50 = R$ 5.95. So each beam will cost 1.92 + 5.95 = R$ 7.87. The total cost of the beams will be

$$19 \cdot 7{,}87 = R\$\ 149{,}53.$$

Solution (item d):

In order to calculate the cost of the floor, we have to consider the area of the circular base of the cistern. Thus, according to Example 01 on page 32, we have the area of the base of the cistern equal to 9.07 m^2 . So the volume of the floor *(V_p)* is

$$V_p = 9{,}07 \cdot 0{,}07 = 0{,}634\, m^3.$$

Thus, 0.634m3 of concrete will be needed to make the floor. So the cost of building the floor is

$$0{,}634 \cdot 240{,}00 = R\$\ 152{,}16$$

Solution (item e):

For the internal wall plaster, we will first calculate the volume of the material used. So the volume of the plaster (V_r) is

$$V_r = 19{,}21 \cdot 0{,}02 = 0{,}384\, m^3$$

This will require 0.384m3 of mortar to build the plaster. We can now estimate the cost of the plaster, which is

$$0{,}384 \cdot 200{,}00 = R\$\ 76{,}80$$

Now we can calculate the total cost of the cistern. This way, we'll add up all the solutions we've found so far, plus the estimated labor cost. So the total cost of the cistern is

$$C_{total} = 113{,}40 + 45{,}60 + 149{,}53 + 152{,}16 + 76{,}80 + 1100{,}00$$
$$= \boldsymbol{R\$\ 1.637{,}49}$$

4.5.2 Problem 02 - How many liters of rainwater can a roof measuring 6 x 9 m collect, assuming that it receives an annual rainfall of 500 mm with a roof efficiency of 80%? And what would be the volume of a slab cistern needed to hold this amount of captured rainwater?

Data:

Roof area: $A_L = 9 \cdot 6 = 54\ m^2$

Annual rainfall: $P_A = 500\text{mm} = 0.50\text{m}$

Capture efficiency: $E_c = 80\% = 0.80$

Volume captured: V, knowing that $Im^3 = 1000$ *liters*

Solution:

$V = At \cdot Pa \cdot Ec$

$V = 54 \cdot 500 \cdot 80 = 54 \cdot 0.5 \cdot 0.8 = 21.6m^3 = 21600$ *liters*

Thus, a roof measuring 54 m^2 , which receives 500 mm of rain with a capture efficiency of 80%, can supply the volume needed for a cistern with a capacity of 21,600 liters. This cistern, with this volume of 21.6 m^3 during the rainy season, supplies a family of 5 people who use about 200 liters a day for 108 days.

4.5.3 Inverse Problem: And what is the area of the roof in m2 to supply a plate cistern with a volume of 16,000 liters?

Data:

A = Roof area

V = volume of cistern = 16,000 *l = 16* m^3

B = annual rainfall = 500 *mm* = 0.50 *m*

$C = 80\%$ *capture efficiency* $= 0.8$

Solution:

$V = A \cdot B \cdot C \rightarrow A = V / (B\text{-}C)$

$B\text{-}C = 0.50 \cdot 0.80 = 0.40$

$A = 16m^3 / 0.40 = 40m^3$

So, to supply a cistern with 16,000 liters of water, all we need is a roof with an area of 40m .[2]

FINAL CONSIDERATIONS

Nonetheless, we cannot fail to report how much this work has had an impact on the relationship between parents, children and others with regard to the teaching and learning of mathematics in the Barra do Japi Community.

It was then through the teachings and techniques of Ethnomathematics that we were able to make a contribution to the Barra do Japi community and the surrounding area by learning mathematical knowledge related to the basic geometry intrinsic to the construction of a slab cistern. This allowed us to study plane figures such as rectangles, triangles and circles, as well as some figures in space such as cylinders and cones. It also allowed us to study the learning related to calculating the length of a circle, the area of a circle, the area of a rectangle, the study of radius, diameter and height. As well as calculating the volume of a cylinder, and other mathematical calculations related to converting rainfall in mm to the volumetric capacity of a cistern.

Furthermore, we also realized that it is through ethnomathematics that we can relate to the community, interacting with everyday practice within a particular activity that we intend to study, in order to make the link between theory and practice. And so we can extract all the knowledge involved, which will allow us to learn more from the authors involved.

So we can say that this work was of great relevance and that it turned out to be very rewarding for those involved. It leaves us with a legacy of the teaching-learning process, the scientific approach to academic art, formality, ethics, perseverance and, why not, hope. Since in this one we had the intention of intervening, interacting and aiming for something.

Finally, we can say with certainty that we have achieved our goal and that: Those who teach learn by teaching. And those who learn teach by learning (Paulo Freire, 2000).

REFERENCES

BRASIL, **National Curriculum Parameters**: mathematics. Brasilia: MEC/SEF, 1997.

BRASIL, **National Curriculum Parameters**: Mathematics. 3.ed. Brasilia: MEC/SEF, 2001.

BRASIL. Law no. 9394, of December 20, 1996. **Law of Guidelines and Bases of Education**. 4. ed. Brasilia: Senado Federal, 2007.

D'AMBROSIO, Ubiratan. **Ethnomathematics: the** link between tradition and modernity. Belo Horizonte: Autêntica, 2001.

D'AMBROSIO, Ubiratan. **Mathematics Education:** from theory to practice. 2.ed. Campinas: Papirus, 1997.

D'AMBROSIO, Ubiratan. **Ethnomathematics:** art or technique of explaining and knowing. Sao Paulo: Atica, 1990.

DOLCE, Osvaldo; POMPEO, José Nicolau. **Fundamentals of elementary mathematics 10**: spatial geometry, position and metrics. 6. ed. Sao Paulo: Atual, 2005.

FREIRE, Paulo. **Educaçâo e Mudança**. 26. ed. Rio de Janeiro: Paz e Terra, 2002.

FREIRE, Paulo. **Pedagogy of Autonomy**. 31. ed. Rio de Janeiro: Paz e Terra, 2000.

FERREIRA, Edson Luiz Cataldo. **Basic geometry**. v.2./3.ed. Rio de Janeiro: Atual, 2007.

LÜDKE, M; ANDRÉ, M. E.D. **Pesquisa em educaçâo:** abordagens qualitativas. Sao

Paulo: EPU, 1986.

MATTOS, J. R. L. & BRITO, M. L. B. **Rural agents and their professional practices:** a link between mathematics and ethnomathematics. Bauru: Ciência & Educaçao, 2012.

MACHADO, S. D. A. **Didactic Engineering.** In: MACHADO, S. D. A. (org). **Educaçâo Matemàtica:** Uma (nova) introduçao. 3. ed. Sao Paulo: EDUC, 2008.

Printed by Books on Demand GmbH, Norderstedt / Germany